FUN FACT FILE:
EARTH
SCIENCE

20 FUN FACTS ABOUT ASTRONOMY

By Jill Keppeler

Gareth Stevens
PUBLISHING

Please visit our website, www.garethstevens.com. For a free color catalog of all our high-quality books, call toll free 1-800-542-2595 or fax 1-877-542-2596.

Cataloging-in-Publication Data
Names: Keppeler, Jill.
Title: 20 fun facts about astronomy / Jill Keppeler.
Description: New York : Gareth Stevens Publishing, 2018. | Series: Fun fact file: earth science | Includes index.
Identifiers: ISBN 9781538211731 (pbk.) | ISBN 9781538211755 (library bound) | ISBN 9781538211748 (6 pack)
Subjects: LCSH: Astronomy–Juvenile literature.
Classification: LCC QB46.K47 2018 | DDC 520–dc23

First Edition

Published in 2018 by
Gareth Stevens Publishing
111 East 14th Street, Suite 349
New York, NY 10003

Copyright © 2018 Gareth Stevens Publishing

Designer: Sam DeMartin
Editor: Joan Stoltman

Photo credits: Cover, pp. 1, 12, 13, 16 Vadim Sadovski/Shutterstock.com; p. 5 peresanz/Shutterstock.com; p. 6 Fotokvadrat/Shutterstock.com; pp. 7, 11 Aphelleon/Shutterstock.com; p. 8 NikoNomad/Shutterstock.com; p. 9 Dotted Yeti/Shutterstock.com; p. 10 Triff/Shutterstock.com; p. 14 edobric/Shutterstock.com; p. 15 (planets) Alexey Yarkin/Shutterstock.com; p. 15 (United States) vectorEPS/Shutterstock.com; p. 17 NASA Images/Shutterstock.com; p. 18 LifetimeStock/Shutterstock.com; p. 19 Jurik Peter/Shutterstock.com; p. 20 Buyenlarge/Archive Photos/Getty Images; p. 21 (all) Sergey Mikhaylov/Shutterstock.com; p. 22 Education Images/UIG/Universal Images Group/Getty Images; p. 23 Bettman/Bettman/Getty Images; p. 24 dmitry_islentev/Shutterstock.com; p. 25 Interim Archives/Archive Photos/Getty Images; p. 26 Sietec/Wikimedia Commons; p. 27 Prof.Professorson/Wikimedia Commons; p. 29 Kostenko Maxim/Shutterstock.com.

Printed in the United States of America

CPSIA compliance information: Batch #CW18GS: For further information contact Gareth Stevens, New York, New York at 1-800-542-2595.

Contents

Words in the glossary appear in **bold** type the first time they are used in the text.

A Whole New World

Have you ever gone outside at night and looked at the sky? On a night without clouds, you can often see stars, the moon, and maybe even another **planet**!

Astronomy is the study of stars, planets, and other things in space. The scientists who study space are called astronomers. Their job is very exciting these days as the tools used to see deep into space get better and better. New discoveries about space happen every year!

We can see many stars just by looking up at the night sky, but scientists use **telescopes** and other tools to see much more than our eyes ever could!

5

FACT 1

When you look at the stars, you're looking back in time.

Stars are so far away that their light takes a long time to reach Earth. You're seeing them the way they looked years ago—sometimes thousands of years ago!

Light from the farthest stars in our own Milky Way **galaxy** can take 100,000 years to reach Earth.

Stars can be different colors.

Stars usually look white to us, but they're often actually different colors. Red stars are very cool. Blue stars are very hot. There are also yellow-orange stars and even green stars!

Earth's sun is a yellow-orange star.

FACT 3

There are billions of planets and stars just in our galaxy.

There are eight planets in our **solar system**. However, experts believe there are at least 100 billion planets in our galaxy—and about 1,500 of them are within 50 **light-years** of Earth.

There are many planets in our galaxy. Scientists think there may be more than 100 billion galaxies in the **universe**!

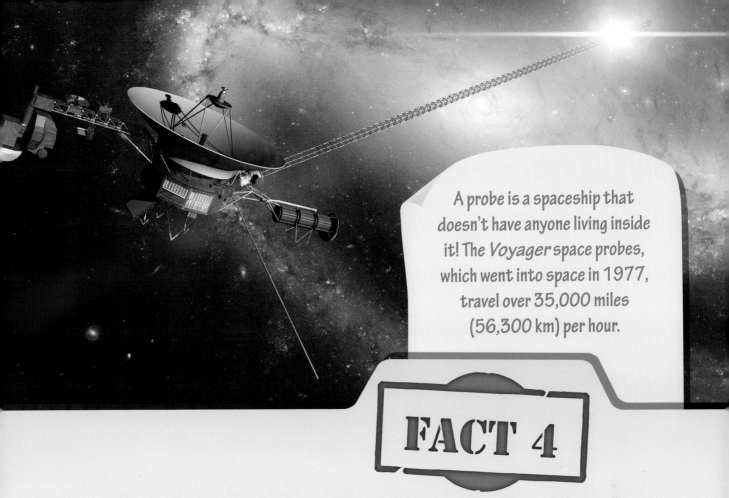

A probe is a spaceship that doesn't have anyone living inside it! The *Voyager* space probes, which went into space in 1977, travel over 35,000 miles (56,300 km) per hour.

FACT 4

It would take a jet about 120 billion years to fly across the Milky Way.

The *Voyager* space probes could cross the Milky Way in *only* 2 billion years. Even if the probes could travel at the speed of light, it would still take 100,000 years!

Earth is over 7,900 miles (12,700 km) wide at the **equator**. Three planets in our solar system—Venus, Mars, and Mercury—are smaller than Earth.

FACT 5

One million Earths could fit inside the sun.

The sun is about 865,000 miles (1,392,000 km) wide, which is actually pretty small for a star. Even though it's small compared to other stars, the sun contains 99.8 percent of the mass of our entire solar system!

FACT 6

The biggest star ever discovered is about 1,700 times as big as the sun.

UY Scuti, a star that's about 9,500 light-years away from Earth, is almost 1.5 billion miles (2.4 billion km) wide!

Our sun may be much smaller than UY Scuti, but it's large enough to provide energy for all the life on Earth!

The planets in our solar system are Mercury, Venus, Earth, Mars, Jupiter, Saturn, Uranus, and Neptune. Mercury is the closest to the sun, and Neptune is the farthest.

FACT 7

There used to be nine planets in our solar system.

If you ask your parents how many planets there are, they might say nine. That's because Pluto was considered a planet until 2006, when scientists decided it was a dwarf, or small, planet.

It's very cold on Pluto:
−387°F (−233°C)!

FACT 8

An 11-year-old girl gave Pluto its name.

After Pluto was discovered in 1930, Venetia Burney of
England suggested that it should be named after the Roman
god Pluto. The scientists who discovered it agreed!

A lot of what we know about Pluto came from the *New Horizons* space probe, which passed within 7,800 miles (12,500 km) of Pluto in 2015.

FACT 9

Pluto is smaller than the United States.

Pluto is only about 1,400 miles (2,250 km) wide. The distance from California to Maine is about 2,900 miles (4,700 km)!

How Small Is Pluto?

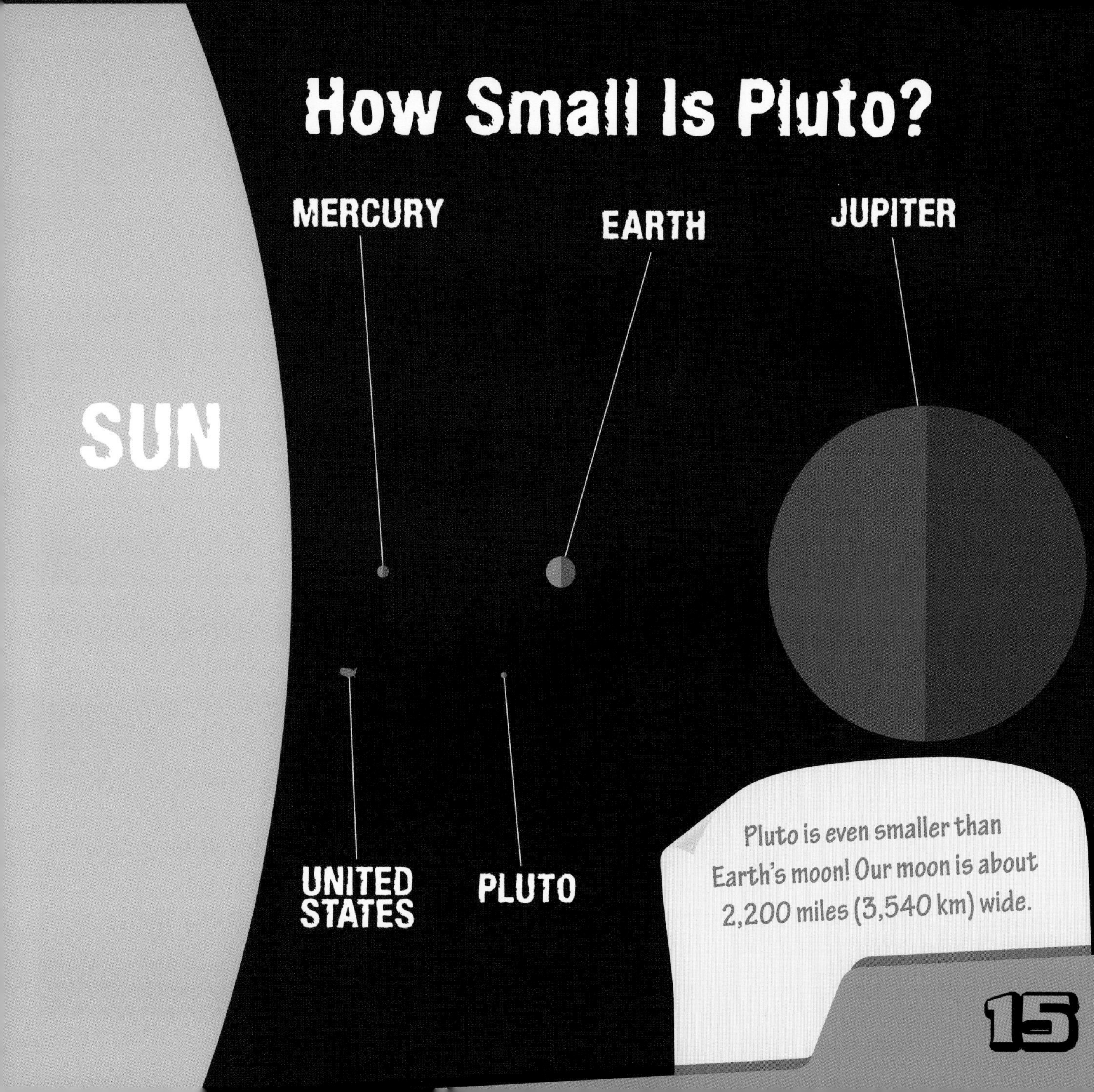

MERCURY

EARTH

JUPITER

SUN

UNITED STATES

PLUTO

Pluto is even smaller than Earth's moon! Our moon is about 2,200 miles (3,540 km) wide.

FACT 10

Uranus is the only planet that rotates on its side.

Uranus rolls over and over from top to bottom rather than spinning round and round. Scientists think something big smashed into the planet a long time ago, tipping it over.

Uranus

A few other planets are also tilted. Even Earth is tilted a little!

Venus

Venus is also the hottest planet in our solar system, even though Mercury is closer to the sun. Venus is about 930°F (500°C)!

FACT 11

Venus, however, spins backward.

Almost all the planets in our solar system spin counterclockwise, or to the left. Venus, however, spins clockwise, or to the right. Scientists think it flipped over at some point.

FACT 12

There's a planet made of diamond.

Scientists are studying a planet in the Milky Way that's made, in part, of the **mineral** diamond. They think the planet, which is named 55 Cancri e, used to be a star.

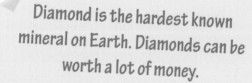

Diamond is the hardest known mineral on Earth. Diamonds can be worth a lot of money.

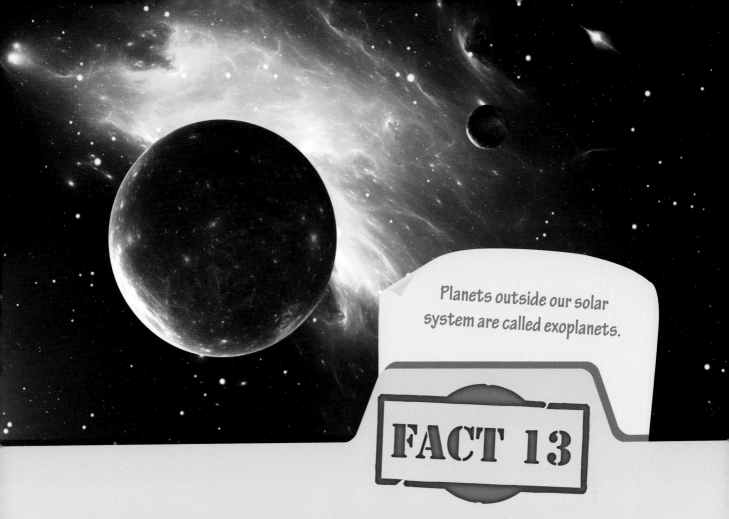

Planets outside our solar system are called exoplanets.

FACT 13

Some planets **orbit** two suns!

It might sound impossible, but scientists have found a few planets outside our solar system that orbit two suns. They even have double sunsets! In fact, one planet has three suns!

FACT 14

There are bears in the sky. Sort of.

A group of stars that forms a shape and has been given a name is called a constellation. Ursa Major and Ursa Minor are constellations shaped like bears!

There are 88 constellations we can see from Earth. Many are shaped like animals or people.

Connect the Dots

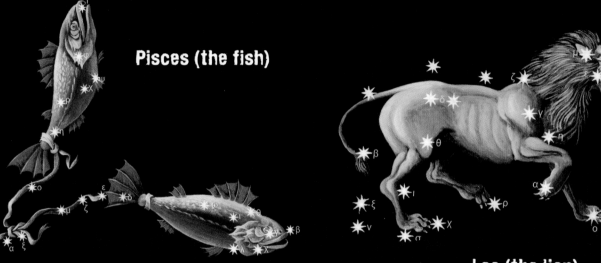

Pisces (the fish)

Leo (the lion)

Draco (the dragon)

The constellations make many different shapes, though sometimes they're hard to make out.

FACT 15

There are rocks from Mars on Earth.

Humans have never visited Mars, so how did its rocks get here? They're meteorites, which are pieces of rocks or metal that have fallen to Earth from outer space.

ALH84001,0

This meteorite from Mars was found in Antarctica! It has since been moved to the Johnson Space Center in Houston, Texas.

Halley's comet, 1986

A comet is an object made of ice and dust that moves through outer space. Comets get a long, bright tail when they pass near the sun.

FACT 16

Halley's comet is going to pass Earth again in 2061.

Every 75 years or so, Halley's comet passes close enough to Earth that we can see it. It last passed in 1986. You'll be a grown-up by the time it returns!

FACT 17

People used to think that the sun moved around Earth.

We know now that the planets in our solar system all move around the sun. Many years ago, people had this mixed up. People didn't believe the first astronomers who said otherwise!

In the early 1500s, astronomer Nicolaus Copernicus made a model of the solar system with the sun in the center.

Maria Mitchell discovered a comet through an ordinary telescope.

Maria Mitchell discovered a comet in 1847. It was even named after her! She later became the first woman in the United States to be an astronomer for a living.

Today, Mitchell's telescope is at the Smithsonian Institution National Museum of American History in Washington, DC.

FACT 19

Animals traveled into space before people.

The US and Russian space programs sent a number of animals, including fruit flies, monkeys, a dog named Laika, and a chimpanzee named Ham into space before sending people.

Ham traveled to space in January 1961 and returned safely. He's shown here prior to lift-off.

26

More than 500 people have been to space since 1961.

In April 1961, Yuri Gagarin of Russia became the first person to travel into space. In May 1961, Alan Shepard became the first American to go to space.

Only 12 people have ever walked on the moon. Neil Armstrong of the United States was the first on July 21, 1969, followed by Buzz Aldrin, shown.

Reach for the Stars!

These fun facts are only a little bit of what we know about space, the stars, and other planets. Scientists continue to study our neighbors in space—both the ones in our solar system and the ones farther away.

Time will tell what new and wonderful things we'll learn about astronomy. Maybe astronomers will discover new planets with life. Maybe humans will be able to visit Mars—or even live there! Maybe you'll be an astronomer and make a great discovery!

Would you like to be an astronaut, or someone who travels in space? Maybe someday you will be!

29

Glossary

equator: an imaginary line around Earth that is the same distance from the North and South Poles

galaxy: a large group of stars, planets, gas, and dust that form a unit within the universe

light-year: the distance light can travel in one year

mineral: matter in the ground that forms rocks

orbit: to travel in a circle or oval around something, or the path used to make that trip

planet: a large, round object in space that travels around a star

rotate: to turn around a fixed point

solar system: the sun and all the space objects that orbit it, including the planets and their moons

telescope: a tool that makes faraway objects look bigger and closer

tilted: slanted, not straight up and down

universe: everything that exists

For More Information

Books

Arlon, Penelope, and Tory Gordon-Harris. *Planets: A LEGO Adventure in the Real World*. New York, NY: Scholastic, 2016.

Carney, Elizabeth. *Planets*. Washington, DC: National Geographic, 2012.

TIME for Kids. *All About Space*. New York, NY: Time Home Entertainment, Inc., 2014.

Websites

Astronomy for Kids
astronomy.com/observing/astro-for-kids
This website has space-themed projects and lots of information about stars and planets.

HubbleSite
hubblesite.org
This site is all about the Hubble Space Telescope, one of the most powerful telescopes!

NASA Kids' Club
nasa.gov/kidsclub/index.html
Play games and learn about space!

Index